# Internet Quests

# WEATHER

## Primary

AF710526

**Author**

*Judy Gaborvec*


**Editor**
Sara Connolly

**Editorial Manager**
Elizabeth Morris, Ph.D.

**Editor-in-Chief**
Sharon Coan, M.S. Ed.

**llustrator**
Bruce Hedges

**Cover Design**
Jeff Sutherland

**Cover Artist**
Jose Tapia

**Photo Credit**
Digital Imagery © copyright 2000 PhotoDisc, Inc.

**Art Coordinator**
Denice Adorno

**Imaging**
Ralph Olmedo, Jr.

**Product Manager**
Phil Garcia

**Published in partnership with**
Ready Ed Publications
11/17 Foley Street
Balcatta WA 6021
Australia
www.readyed.com.au

**Publishers**
Rachelle Cracchiolo, M.S. Ed.
Mary Dupuy Smith, M.S. Ed.

Teacher Created Materials

***Teacher Created Materials, Inc.***
6421 Industry Way
Westminster, CA 92683
www.teachercreated.com

**ISBN-0-7439-3407-5**



*©2001 TeachOer Created Materials, Inc.*



Reprinted, 2001

Made in U.S.A.



**The classroom teacher may reproduce copies of materials in this book for classroom use only. The reproduction of any part for an entire school or school system is strictly prohibited. No part of this publication may be transmitted, stored, or recorded in any form without written permission from the publisher.**

# Table of Contents

# Introduction

## How To Use This Book

### 1. Internet Activity Pages

**Note:** As this is not an introductory book on using the Internet, you will need to have some understanding of how to navigate the World Wide Web and how to open various Internet addresses prior to attempting these Internet-related activities.

To make use of the pages headed **Internet Activity Pages**, access the Teacher Created Materials World Wide Web site at the following address:

**http://www.teachercreated.com/books/3407**

(Bookmark this site for easy access later.)

When the page opens in your browser, you will see a listing of page numbers and links to Web site addresses. If you click on one of the provided links, you will be taken to the Web page without having to laboriously type the address.

The CD-ROM included with this book also contains hyperlinks that will automatically connect you to the Web sites on the Internet.

### 2. Alternet Activity Pages

To make use of the pages headed **Alternet Activity Pages**, insert the Alternet CD into your computer's CD-ROM drive. The HTML files on the Alternet CD-ROM require either *Microsoft Internet Explorer* or *Netscape Navigator* to run, so you will need to access one or the other of these browsers. The CD-ROM contains the Alternet package relevant to this book.

The minimum hardware requirement for use of the Alternet program is any computer that uses a frame-capable browser. This includes both Macintosh® and Windows® operating systems.

### 3. How To Use the Alternet

The Alternet CD can be used on both IBM compatible and Macintosh computers. Although you will need a browser to view the files, the computer does not need to be connected to the Internet as all files are self-contained.

You may use the files directly from the CD by clicking on the *Weather* folder and then opening the file [*begin.htm*]. For faster operation you can copy the contents of the *Weather* folder to a folder on your hard disk, remove the CD for safekeeping, then open [*begin.htm*] on your hard disk.

It is recommended that you create a new folder for the Alternet set for each book you have purchased. (*Plants*, *Animals*, etc.) Name the folder after the book you are using.

**Note:** When you load the Alternet program for your students to use, the browser may give warning messages about connection difficulties. This is because you are not attempting to connect to the Internet. You can ignore this by clicking **OK**.

**Note:** Purchase of a single book gives you the right to copy the Alternet program on up to ten computers.

# Teachers' Notes

## Teacher Information

This series is designed to provide teachers of primary-age children with a multimedia-based set of resources for the Weather theme.

In using this multimedia resource, students will be required to learn and utilize a range of associated skills such as:

- reading and interpreting given information.
- summarizing given information.
- using generalizations based on given information.
- navigating their way around a Web site.

## What does this multimedia approach consist of?

This series aims to provide teachers with a three-pronged approach to meeting the requirements of concepts related to the scientific research curriculum. It combines the use of traditional information sources such as reference books and materials with the enormous information capabilities of the Internet and the appeal of computer databases as provided by our Alternet program. As many schools face practical limitations in trying to make use of information accessed through the Internet and computer software, the materials provided here are aimed at having the class working as three groups, each using different methods of information retrieval, which eases the pressures on the available hardware and facilities.

1. **The Internet**

   The use of the related activity pages in this book (titled **Internet Activity Pages**) together with the information sources provided by the World Wide Web lends a contemporary and exciting aspect to studies in this subject area. As detailed on Page 2 of this book, students access the relevant Web pages through the Teacher Created Materials Web site and use the information they find to complete the associated tasks on the activity page.

2. **The Alternet Program**

   The Alternet is a database of information and pictures related to the strand of study on which students are working. Its appearance emulates Web sites found on the Internet's World Wide Web and it is navigated in exactly the same way as the Web through the use of browsers. As with the Internet activities described above, students use the information gathered from the files to complete the relevant activity pages in this book (titled **Alternet Activity Pages**).

3. **General Activities**

   For the activity pages in these sections (titled **General Activity Pages**), students utilize traditional print-related reference materials to complete the tasks on the sheets. It would be useful for a collection of appropriate books to be assembled before commencing the unit so these can be accessed and used with as little disruption as possible. Alternatively, Web site references have been included where possible to provide background information for teachers.

 *© Teacher Created Materials, Inc.*

# Teachers' Notes *(cont.)*

## Updating and Checking of Internet Addresses

The dynamic nature of the Internet means that some sites may change URLs or even disappear altogether. An ongoing role of the publishers will be to monitor changes and post them on our Web site. All addresses used are checked weekly and changes will be posted on our Web site.
**http://www.teachercreated.com/books/3407**

## Suggested Use of This Series

As indicated, a multimedia approach has been utilized not only to provide students with a range of reference sources but also to take account of the difficulties schools have in trying to match limited resource access with the requirements of the curriculum. To help teachers plan for these contingencies, this book has been structured to allow units of work where groups rotate from one information source to the next on a week to week basis.

It is important to note that the activities in each section are "stand-alone," meaning that students will not need to complete previous activities in order to work successfully. Exactly how teachers choose to use the materials is obviously dependent on the amount of computer hardware and Internet access available, but with a bare minimum in mind, books have been designed to allow a rotation (assuming one period per week is spent on the unit).

To help follow a planning pattern such as this, the following checklist can be used to check off the activity sheets that each group has completed:

1. Where possible, units are evenly balanced as in the example below, i.e., there is an equivalent number of Internet, Alternet, and General activities in each section of the book to allow for easy rotation of groups.
2. Activity pages are designed to take about 25–30 minutes—the approximate length of time usually available in a school period after preliminaries have been sorted out.

## Check List

| Internet | Groups 1 | 2 | 3 |
|---|---|---|---|
| Snowflakes | | | |
| Lightning Science | | | |
| Lightning | | | |
| Clouds All Around | | | |
| Cloud Shapes | | | |
| Tornado Talk | | | |
| Tornado Safety Tips | | | |
| Weather Conditions | | | |
| Weather Warning! | | | |
| Weather Instruments | | | |
| Online Weather Games | | | |

| Alternet | Groups 1 | 2 | 3 |
|---|---|---|---|
| What Makes the Weather? | | | |
| Weather Forecasting | | | |
| Measuring the Weather | | | |
| Water | | | |
| The Water Cycle | | | |
| The Wind | | | |
| The Sun | | | |
| Desert Weather | | | |
| Arctic Weather | | | |
| Tropical Rain Forest | | | |
| The Seasons | | | |

| General | Groups 1 | 2 | 3 |
|---|---|---|---|
| Let's Make a Cloud! | | | |
| Super Snow | | | |
| Measuring the Wind 1 | | | |
| Measuring the Wind 2 | | | |
| Wind Power | | | |
| Hot and Cold | | | |
| Water, Water, Water | | | |
| Let's Make Rain! | | | |
| Rainbows | | | |
| | | | |
| | | | |

# Weather Phenomena

## Content Area(s):

- earth science

## Objectives:

In this section students:

- describe how snowflakes are formed.
- explain how lightning occurs.
- identify snowflake shapes.
- investigate lightning facts.
- investigate snow facts.
- navigate their way around a Web site.

## Materials Required:

- computer with Internet access
- pencil or pen

## Web Site(s):

- Students need to access the TCM web page for the Weather book at:

  **http://www.teachercreated.com/books/3407**

  They then click on the link to the page they are working on.
- Let the children familiarize themselves with the Web site before starting the activity sheets.

## Time:

- approximately 20–35 minutes

## Background Information:

### Snowflakes (Page 7)

- Snowflakes start as tiny ice crystals that join up with other crystals to form snowflakes.
- Snow forms when the temperature on the top of the clouds is below freezing.

### Lightning Science (Page 8)

- Lightning occurs when electricity builds up in a thunder cloud. The clouds become negatively charged and a brilliant flash of light will hit the closest positively charged object. A stroke of lightning then zig-zags to the ground (or to another cloud). The lightning we see is the "return" stroke that races back up to the clouds. Lightning can travel from clouds to the ground or from cloud to cloud.

### Lightning (Page 9)

- Lightning and thunder occur at the same time but you see the lightning first because light travels faster than sound. If you hear a thunderclap five seconds after you see a flash, the storm is about one mile or 1.6 kilometers away.
- Lightning always hits the tallest positively charged object because it always takes the shortest pathway.
- It is possible to survive a lightning strike. Lightning can strike in the same place twice.

 *© Teacher Created Materials, Inc.*

Name: ______________________________

# Snowflakes

Go to **http://www.teachercreated.com/books/3407**

Click on **Page 7**, **Site 1**

❄ Describe how snowflakes form ____________________

____________________________________________

____________________________________________

____________________________________________

____________________________________________

❄ Draw four of the different kinds of snowflake shapes in the boxes below.

| | |
|---|---|
| | |
| | |

## Snow Facts—True or False?

1. Snow can act like a mirror and reflect heat. ____________________
2. Snow will form if the temperature on the top of the clouds is 50° F or 10° C. ____________
3. It can only snow if there are clouds in the sky. ____________________
4. Snowflakes always have six sides. ____________________
5. There have never been two snowflakes found that are identical. __________
6. 0°Celsius and 32°Fahrenheit are the same temperature. __________

Name:______________________________

# Lightning Science

Go to **http://www.teachercreated.com/books/3407**

Click on **Page 8**, **Site 1**

Lightning is a giant electrical spark in the sky which mostly takes place between a cloud and the ground. When lightning occurs in the atmosphere, its electrical energy scatters in the air.

❄ In the space below, explain in your own words how lightning occurs. Be sure to include the different types of lightning bolts. Illustrate your work with diagrams.

 © Teacher Created Materials, Inc.

Name: ______________________________

# Lightning

Go to **http://www.teachercreated.com/books/3407**

Click on **Page 9**, **Site 1**

❄ Read the paragraph "Before a storm." Fill in the missing information.

Keep an eye on the ________________. On a hot day, ________________ clouds build up. They grow ________________, towering higher and ________________. They________________. The top of the clouds may be spread out by the ________________ at high altitudes. As clouds develop, ________________ in the clouds become ________________ with ________________. When the charges ________________, they make a lightning flash. Lightning may go from one part of a ________________ to another, or from cloud to________________.

It may go from a cloud to ________________ or from earth to a cloud. Lightning takes the ________________ path, so it hits the ________________ objects—a tall tree or house, a tower, or a person standing alone in a ________________.

❄ Why does lightning always hit the highest objects?

______________________________________________________________

______________________________________________________________

❄ Click on the other links on this page to find out if these statements about lightning are true or false.

Lightning and thunder occur at the same time. __________

It is impossible to survive if you are struck by lightning. __________

Lightning never strikes in the same place twice. __________

# Clouds

## Content Area(s):

- earth science

## Objectives:

In this section students:

- discover what clouds are made of.
- investigate different types of clouds.
- discuss why clouds change shape.
- observe clouds in the sky.
- relate weather to cloud type.
- navigate their way around a Web site.

## Materials Required:

- computer with Internet access
- pencil or pen
- Internet Activity Pages

## Web Site(s):

- Students need to access the following page on the Teacher Created Materials Web site:

**http://www.teachercreated.com/books/3407**

They then click on the link underneath the number of the page they are working on. In some cases, there will be more than one Web site for a page, and students will need to click on the links under "Site 1" or "Site 2."

- Allow students to familiarize themselves with the Web site before they begin working on their activity sheets.

## Time:

- approximately 20–35 minutes.

## Background Information:

### Clouds All Around (Page 11)

- Clouds are formed when air rises due to evaporation of water by the sun. As the air rises it cools and the vapor changes back into tiny drops of water. The drops group together and form clouds.
- There are three main types of clouds—***cirrus*** (curl), ***cumulus*** (heap), and ***stratus*** (layer). *Cirrus* clouds look like wispy threads and float high in the sky. They can mean that unsettled weather is on the way. *Cumulus* clouds are large fluffy clouds that usually mean fine weather but if a violent uprush of air forms cumulus clouds then the clouds turn black and thunder, lightning and rain may occur. *Stratus* clouds are low gray layers which can mean that heavy snow or rain is likely. *Alto*, which means high, and *nimbus*, which means rain-bearing, are also used to describe clouds. There are seven other cloud types—cirrocumulus, cirrostratus, altostratus, altocumulus, nimbostratus, stratocumulus, and cumulonimbus.

### Cloud Shapes (Page 12)

- Cloud watching is called *nephelococcygia.*
- Clouds change shape because of the constant movement of the air.

 © Teacher Created Materials, Inc.

Name: ______________________________

# Clouds All Around

Go to **http://www.teachercreated.com/books/3407**

Click on **Page 11**, **Site 1**

**Observing Clouds**

❄ What are clouds made of? ____________________________________________

__________________________________________________________________

__________________________________________________________________

❄ What are the three main types of clouds? Name and describe them below.

1. ______________________ — ____________________________________

   ____________________________________________________________

2. ______________________ — ____________________________________

   ____________________________________________________________

3. ______________________ — ____________________________________

   ____________________________________________________________

❄ What type of cloud can you see in the sky today? ______________________

__________________________________________________________________

❄ What type of weather do you think will follow? _________________________

__________________________________________________________________

❄ Draw the clouds that you see.

Name:______________________________

# Cloud Shapes

Go to **http://www.teachercreated.com/books/3407**

Click on **Page 12, Site 1**

**Cloud Cuckooland**

❄ Look at the picture of the clouds at the site. Draw the five hidden animals below. Click on **ANSWERS** to make sure you drew the correct animals.

| | | | | |
|---|---|---|---|---|
| | | | | |

❄ What unusual word describes what you are doing if you are imagining shapes in the clouds?

______________________________

❄ Choose a cloudy day. List all the shapes you can see in the clouds.

______________________________

______________________________

______________________________

❄ Illustrate your favorite cloud shapes.

❄ Why do you think the shapes of the clouds change?

______________________________

______________________________

______________________________

 © Teacher Created Materials, Inc.

# Tornadoes

## Content Area(s):

- earth science

## Objectives:

In this section students:

- learn about the properties of a tornado.
- list the features of different tornadoes.
- learn about tornado safety tips.
- investigate different tornado categories.
- categorize tornadoes using features.
- navigate their way around a Web site.

## Materials Required:

- computer with Internet access
- pencil or pen
- Internet Activity Pages

## Web Site(s):

- Students need to access the following page on the Teacher Created Materials Web site:

  **http://www.teachercreated.com/books/3407**

  They then click on the link underneath the number of the page they are working on. In some cases, there will be more than one Web site for a page, and students will need to click on the links under "Site 1" or "Site 2."
- Allow students to familiarize themselves with the Web site before they begin working on their activity sheets.

## Time:

- approximately 20–35 minutes

## Background Information:

### Tornado Talk (Page 14)

- A tornado is defined as a violently rotating column of air in which winds can reach up to 500 mph. At sea, they become waterspouts.
- Tornadoes have three categories:
  **Weak tornadoes** have a thin rope-like appearance. The wind speed is no greater than 110 mph.
  **Strong tornadoes** have a funnel-shaped cloud associated with the whirling updraft. The rotating wind speed will vary from 110 to 200 mph.
  **Violent tornadoes** have a huge circular cloud wall above the tornado. The wind speed is in excess of 200 mph.

### Tornado Safety Tips (Page 15)

- If there is a Tornado Watch current, the National Weather Service will broadcast up-to-date information on the radio. If you hear the Warning, steps need to be taken to protect yourself. It depends where you are as to what steps you should take. More information is provided on the Answer Page at the Web site on what you can do to protect yourself at home, in a shopping mall, outside at school, or if you live in a trailer.

Name:____________________________

# Tornado Talk

First go to **http://www.teachercreated.com/books/3407**

Click on **Page 14**, **Site 1**

Answer the following using information at the Web site.

- What is a tornado? __________________________________________________________________________________________________________________________________________

- What are the three categories of tornadoes? Give a brief explanation of each type.

  1. ________________ — __________________________
  2. ________________ — __________________________
  3. ________________ — __________________________

Click on **Site 2** for some photos of tornadoes.

- Draw pictures below to illustrate each type of tornado.

| **Weak** | **Strong** | **Violent** |
|---|---|---|
| | | |

 *© Teacher Created Materials, Inc.*

Name: ______________________________

# Tornado Safety Tips

First go to **http://www.teachercreated.com/books/3407**

Click on **Page 15**, **Site 1**

❄ What should you do if a tornado hits and you are . . .

at home? ______________________________

in a shopping mall? ______________________________

outside? ______________________________

at school? ______________________________

in a trailer? ______________________________

❄ Draw a picture to show how you can stay safe during a tornado.

# Predicting the Weather

## Content Area(s):

- earth science

## Objectives:

In this section students:

- explore the current weather conditions where they live.
- discover what weather conditions are considered dangerous.
- find out what measures can be taken to protect themselves if there is an active weather warning current.
- investigate the different instruments used to measure weather conditions.
- navigate their way around a Web site.

## Materials Required:

- computer with Internet access
- pencil or pen
- Internet Activity Pages

## Web Site(s):

- Students need to access the following page on the Teacher Created Materials Web site:

  **http://www.teachercreated.com/books/3407**

  They then click on the link underneath the number of the page they are working on. In some cases, there will be more than one Web site for a page, and students will need to click on the links under "Site 1" or "Site 2."
- Allow students to familiarize themselves with the Web site before they begin working on their activity sheets.

## Time:

- approximately 20–35 minutes.

## Background Information:

### Weather Conditions (Page 17) and Weather Warning! (Page 18)

- The information on the Web site used for pages 17 and 18 is updated every minute so it is accurate and current. The weather conditions that are monitored by the National Weather Service only appear on this site if there is a current active warning.

### Weather Instruments (Page 19)

- Refer to the Answer Page.

### Online Weather Games (Page 20)

- There is a variety of weather related games for the children to explore at this site. The games have been developed by the Federal Emergency Management Agency. More games can be found at these Web sites: **http://www.fema.gov/kids/little.htm** and **http://www.fema.gov/kids/games1.htm.**

 © Teacher Created Materials, Inc.

Name: ______________________________

# Weather Conditions

First go to **http://www.teachercreated.com/books/3407**

Click on **Page 17**, **Site 1**

❄ Click on the state that you live in. Choose the town or city that you live in or the one which is closest to where you live and fill in the following information.

State: ______________________________

Town: ______________________________

Date and time: ______________________________

Wind: ______________________________

Visibility: ______________________________

Sky conditions: ______________________________

Temperature: ______________________________

Dew point: ______________________________

Relative humidity: ______________________________

What was the highest temperature in the last 24 hours? ______________

What was the lowest temperature in the last 24 hours? ______________

❄ Click on **Satellite** and study the satellite map of the United States.

Describe the conditions of the clouds over the state in which you live.

______________________________________________

______________________________________________

______________________________________________

Name:________________________________

# Weather Warning!

Go to **http://www.teachercreated.com/books/3407**

Click on **Page 18**, **Site 1**

This National Weather Service site is updated every minute so people all over the country can have current information about any active weather warning.

1. List the types of weather conditions you can learn about at this site.

   ________________________________________________

   ________________________________________________

2. Write the name of a state that has an **active warning** for each of the weather conditions below.

   Hurricane ________________________________

   Special Marine ________________________________

   Severe Thunderstorm ________________________________

   Tornado ________________________________

3. Select one state that has an active warning. Find out the following information about the weather conditions of that state.

   Active warning ________________________________

   Towns or cities that will be affected ________________________________

   ________________________________________________

   Expected weather conditions ________________________________

   ________________________________________________

   What people in the area can do to protect themselves ________________________

   ________________________________

   ________________________________

 © Teacher Created Materials, Inc.

Name: ______________________________

# Weather Instruments

Go to **http://www.teachercreated.com/books/3407**

Click on **Page 19**, **Site 1**

❄ Read the information about the different ways to measure weather conditions on the site. Answer the following questions.

1. What are the two different ways that the temperature can be measured?

_______________________________________________

2. What does a barometer measure? ______________________

_______________________________________________

3. What three instruments are used to measure how much water is in the air?

_______________________________________

_______________________________________

4. What is "dew point"? _________________

_______________________________________

_______________________________________

5. What instrument is used to measure the direction of the wind? _______________

_______________________________________________

6. What instrument is used to measure wind speed? ______________

_______________________________________________

_______________________________________________

7. How is wind direction measured? ____________

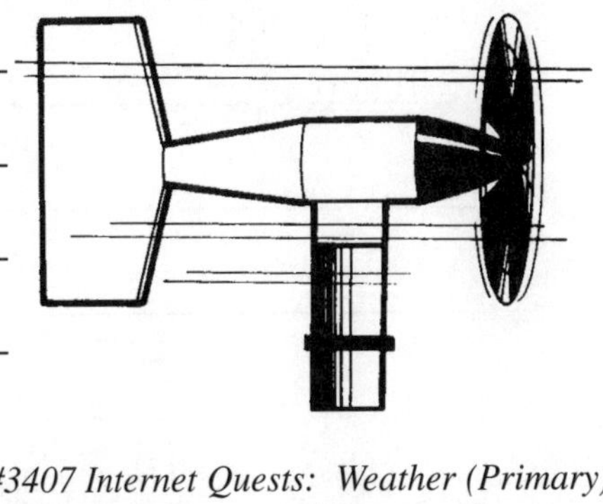

_______________________________________

_______________________________________

_______________________________________

Name:______________________

# Online Weather Games

Go to **http://www.teachercreated.com/books/3407**

Click on **Page 20**, **Site 1**

- Click on **Snowman Games** at Site 1 and experiment with creating different snowmen.

  Draw your favorite snowman creation below and complete the winter landscape.

- Go to **Site 2**. Record the answers to **Do the math to crack the disaster code!** below.

  ______________________

  ______________________

- Go to **Site 3** and read the weather jokes. Write your own weather jokes or riddles below. Submit the best joke by clicking on **Submit a Weather-Related Joke of Your Own**.

  Draw a picture to illustrate your joke.

  ______________________

  ______________________

  ______________________

  ______________________

  ______________________

 © Teacher Created Materials, Inc.

# Weather Components

## Content Area(s):

- earth science

## Objectives:

In this section students:

- identify why weather is important to people.
- investigate weather symbols.
- identify different weather instruments
- discover the four components of weather.
- design their own weather symbols.
- navigate their way around the Alternet.

## Materials Required:

- computer with Internet access
- pencil or pen
- Internet Activity Pages

## Alternet:

- To use the Alternet CD-ROM, open [*begin.htm*]. For more instructions see **How to Use the Alternet** on page 3.
- To go to the individual pages, students access the index page and then click on the number of the page they are using.

## Time:

- approximately 20–35 minutes

## Background Information:

### What Makes the Weather? (Page 22)

- Weather is very important to us because it can change many things about our lives—the clothes we wear, the way we travel, the way we spend our leisure time, and even whether or not we can work. Sometimes the weather can be so bad that people's lives are in danger.
- Weather is made up of four things: temperature, air pressure, movement, and moisture.
- Air pressure is the weight of the air on the earth.
- The wind direction is always measured by the direction the wind is coming from.

### Weather Forecasting (Page 23) and Measuring the Weather (Page 24)

- Weather forecasting is important to people because bad weather conditions can cause harm to people, plants, and animals through storms, hurricanes, snowfalls, blizzards, and other severe weather patterns.
- Weather forecasting and reporting can help people prepare for bad weather conditions.
- Weather patterns are recorded using symbols that can show air pressure (isobars), cold and warm fronts, wind strength and direction, cloud cover, and the likelihood of rain or snow.
- Meteorologists study the weather using a range of instruments ranging from simple thermometers to sophisticated satellite image scanners and radar.

Name: ______________________

# What Makes the Weather?

For this activity you will need the Weather Alternet CD-ROM.

Click on the link to **Page 22**.

❄ Give five reasons why weather is so important to people.

1. ______________________
2. ______________________
3. ______________________
4. ______________________
5. ______________________

❄ Name and explain the four things that make up the weather.

1. ____________ — ______________________
2. ____________ — ______________________
3. ____________ — ______________________
4. ____________ — ______________________

## Weather All Around

❄ Describe the weather around you right now.

______________________

❄ In the box draw an activity that you could do outside in this weather.

 © Teacher Created Materials, Inc.

Name: ______________________________

# Weather Forecasting

For this activity you will need the Weather Alternet CD-ROM.

Click on the link to **Page 23**.

- What do meteorologists do? ________________________________________
  ____________________________________________________________________
  ____________________________________________________________________
- Why is weather forecasting important to people? ______________________
  ____________________________________________________________________
  ____________________________________________________________________
  ____________________________________________________________________
- Weather patterns are recorded using symbols. What do you think these symbols mean?

____________________ ____________________ ____________________

- Design your own symbols for these weather conditions.

| Windy | Hailstones | Thunder and Lightning |
|---|---|---|
| Cloudy | Snowing | Tornado |

Name: ______________________________

# Measuring the Weather

For this activity you will need the Weather Alternet CD-ROM.

Click on the link to **Page 24**.

❄ Look at the weather instruments below. Label each one and describe what it measures.

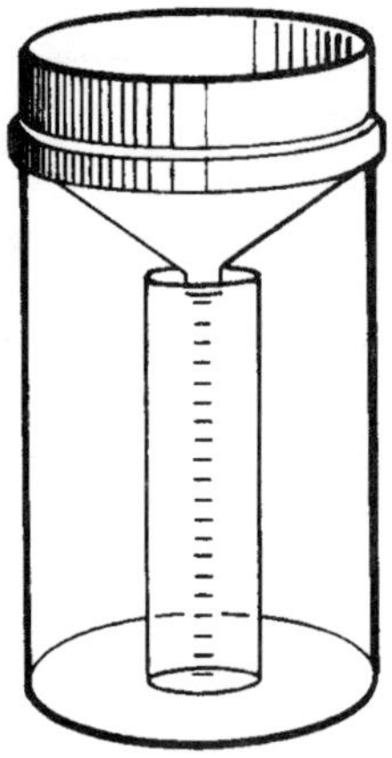

Name: ____________________________

What does it measure? _____________

____________________________________

Name: ____________________________

What does it measure? _____________

____________________________________

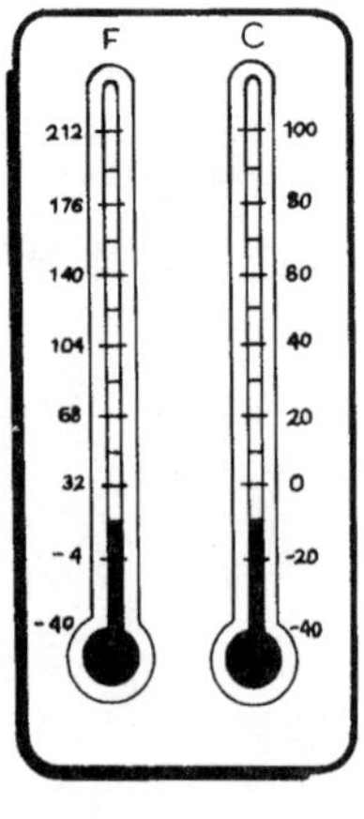


Name: ____________________________

What does it measure? _____________

____________________________________

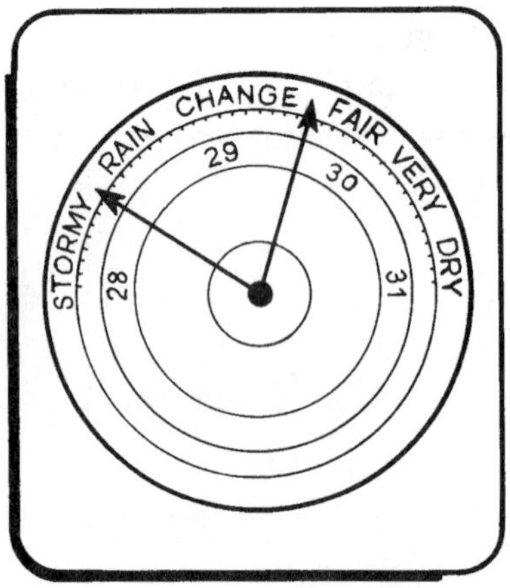


Name: ____________________________

What does it measure? _____________

____________________________________

 © Teacher Created Materials, Inc.

# The Water Cycle

## Content Area(s):

- earth science

## Objectives:

In this section students:

- explore the properties of water.
- investigate wind and its properties.
- discover why the sun is so important to the earth.
- label a diagram of the water cycle.
- describe the different instruments that measure wind.

## Materials Required:

- computer with Internet access
- pencil or pen
- Alternet Activity Pages

## Alternet:

- To use the Alternet CD-ROM, open [*begin.htm*]. For more instructions see **How to Use the Alternet** on page 3.
- To go to the individual pages, students access the index page and then click on the number of the page they are using.

## Time:

- approximately 20–35 minutes

## Background Information:

### Water (Page 26) and The Water Cycle (Page 27)

- Water in the air exists in three forms—water vapor, cloud droplets, and liquid rain drops.

### The Wind (Page 28)

- Wind is moving air that blows because air is always moving from high pressure to low pressure areas.
- An anemometer measures how hard the wind blows. It is sometimes called a wind gauge.
- A meteorologist is a person who studies and forecasts the weather.
- A wind vane measures the direction of the wind. Another instrument that also measures the wind direction is a windsock. The direction of the wind is always measured from where the wind comes from and not the direction to which it is blowing.

### The Sun (Page 29)

- The energy from the sun helps plants grow, and affects the winds, ocean currents, and the water cycle. It also heats up the earth.
- Different parts of the earth heat up differently because of the distance from the sun, the type of surface, and the concentration of the sun's rays on the earth.
- A simple experiment to discover which type of surface in your school yard absorbs the most heat would be to place identical glass containers with a small amount of water in them at different locations around the school. The one with the most water evaporated absorbs the most heat.

Name:______________________________

# Water

For this activity you will need the Weather Alternet CD-ROM.

Click on the link to **Page 26**.

❄ Describe the three different ways that water exists in the air.

1. ______________________________
2. ______________________________
3. ______________________________

❄ Draw a rainbow scene.

❄ Draw a line to match the words to the correct meaning.

| | |
|---|---|
| precipitation | layer of air that surrounds the earth |
| sleet | 32°F or 0°C |
| atmosphere | gas |
| freezing point | amount of water in the air |
| humidity | pull to the earth |
| vapor | rain |
| gravity | rain that freezes on the way to the earth |

 *© Teacher Created Materials, Inc.*

Name: ______________________________

# The Water Cycle

For this activity you will need the Weather Alternet CD-ROM.

Click on the link to **Page 27**.

❄ Fill in the words to tell the story of the water cycle.

1. The ______________________________ dries up ______________________________ from lakes and oceans.
2. The cool air changes the water vapor into tiny droplets of floating ______________________________.
3. The droplets crowd together and form a ______________________________.
4. The wind blows the ______________________________ towards the land.
5. The tiny droplets join together and fall to the ______________________________.
6. The rain soaks into the ground and collects in ______________________________.
7. The ______________________________ that never ends has started again.

❄ Use the Alternet information to label the diagram of the water cycle below.

Name:______________________________

# The Wind

For this activity you will need the Weather Alternet CD-ROM.

Click on the link to **Page 28**.

❄ What is wind and why does it blow?

________________________________________

________________________________________

________________________________________

❄ Draw a windy scene below showing the wind coming from the east.

❄ Find out what these weather words mean.

anemometer: ________________________________________

________________________________________

meteorologist: ________________________________________

________________________________________

wind vane: ________________________________________

________________________________________

windsock: ________________________________________

________________________________________

wind direction: ________________________________________

________________________________________

 © Teacher Created Materials, Inc.

Name: ______________________________

# The Sun

For this activity you will need the Weather Alternet CD-ROM.

Click on the link to **Page 29**.

❄ Give five reasons why the sun is important to the earth.

______________________________

______________________________

______________________________

❄ Why do different parts of the world heat up differently?

1. ______________________________
2. ______________________________
3. ______________________________

❄ Look at the diagram below. Use the words below to label the diagram.

**sun, sun's rays, equator, North Pole, South Pole**

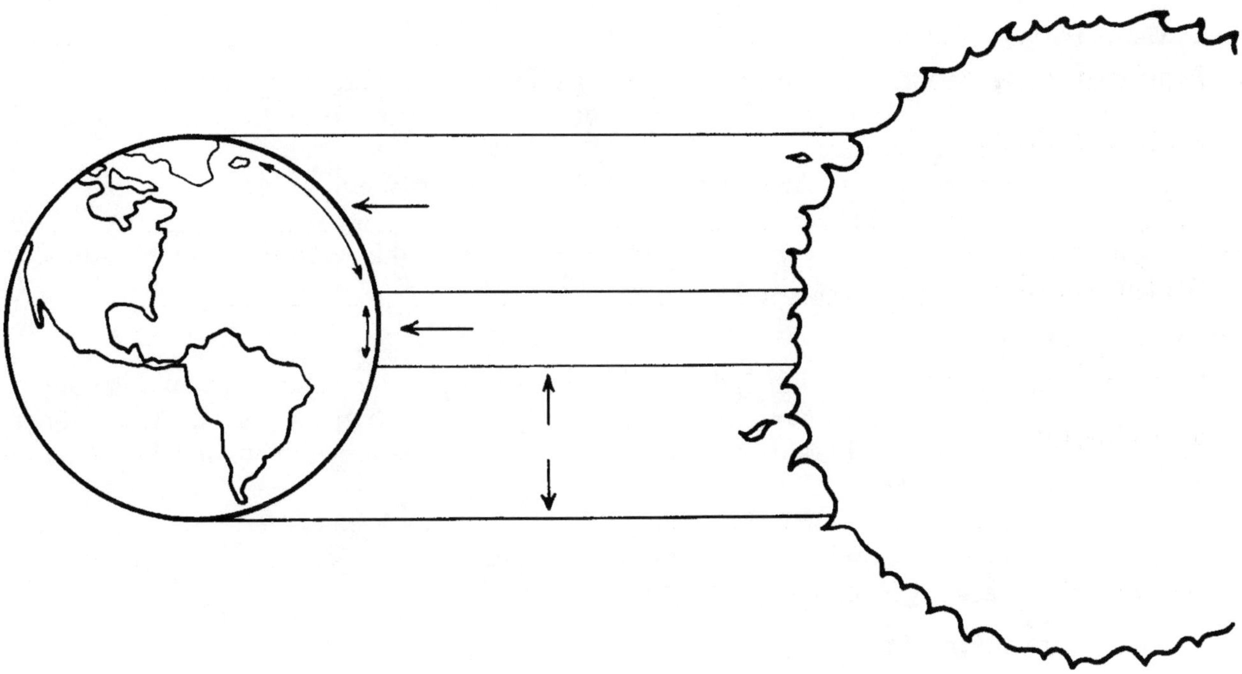

❄ Color the areas of the world which you think are the coldest in blue. Color the hottest areas in red.

# Climates

## Content Area(s):

- earth science

## Objectives:

In this section students:

- investigate the features of a desert.
- investigate tropical rain forests.
- gather facts about the different seasons.
- investigate Arctic weather.
- discover why there are different seasons.
- navigate their way around the Alternet.

## Materials Required:

- computer with Internet access
- pencil or pen
- Alternet Activity Pages

## Alternet:

- To use the Alternet CD-ROM, open [*begin.htm*]. For more instructions see **How to Use the Alternet** on page 3.
- To go to the individual pages, students access the index page and then click on the number of the page they are using.

## Time:

- approximately 20–35 minutes

## Background Information:

### Desert Weather (Page 31)

- Some deserts are hot and some are cold, depending on what part of the world they are in. Some are rocky or sandy or covered in stones. Some have very little plant life and others are covered in dry-looking plants.
- The hottest deserts are close to the middle of the earth (equator) where the sun's rays are strongest and most concentrated.
- Even in the driest desert there is water trapped underground. If this water comes to the surface through a crack in a rock, it is called an oasis.

### Arctic Weather (Page 32)

- In much of the Arctic the earth, ice, and rock are permanently frozen solid. The solid mass is called permafrost. It is covered with a layer of ice and snow which melts in summer. In winter the permafrost can be up to 1,000 feet (300 meters) deep. In summer, it thaws and can be 2 to 6 feet (0.6 to 2 meters) deep.
- The sun gives off less heat at the Arctic circle than at the equator because the sun's rays are less concentrated. Antarctica would experience similar weather conditions while the area around the center of the earth would experience the hottest weather.

### Tropical Rain Forest (Page 33)

- Refer to Answer Page.

### The Seasons (Page 34)

- Review the different hemispheres and why the seasons occur at different times.

 *© Teacher Created Materials, Inc.*

Name: ______________________________

# Desert Weather

For this activity you will need the Weather Alternet CD-ROM.

Click on the link to **Page 31**.

- There are many different types of desert. List some words to describe deserts.

______________________________

______________________________

______________________________

______________________________

______________________________

______________________________

- Find answers to these questions.

1. How much rainfall does a desert area average per year? ______________
2. In what part of the world are the hottest deserts found? ______________

   ______________________________
3. Why are the deserts in this part of the world so hot? ______________

   ______________________________
4. What is an oasis? Draw an oasis in the box.

______________________________

______________________________

______________________________

______________________________

______________________________

______________________________

______________________________

Name:__________________________

# Arctic Weather

For this activity you will need the Weather Alternet CD-ROM.

Click on the link to **Page 32**.

❄ The Arctic Circle is made up of permafrost. Find out these facts about permafrost.

1. What is permafrost?________________________
2. How deep is the permafrost in winter? __________________
3. What happens to the permafrost in summer? _______________
4. How deep is the permafrost in summer?__________________

❄ Look at the diagrams below of how the sun's rays hit the earth. Locate the Arctic Circle on both of the diagrams.

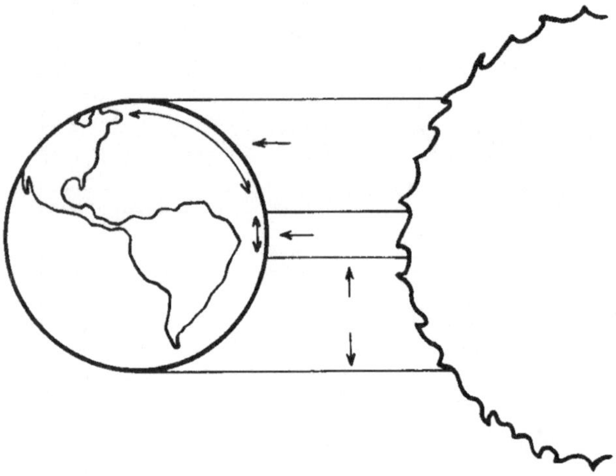

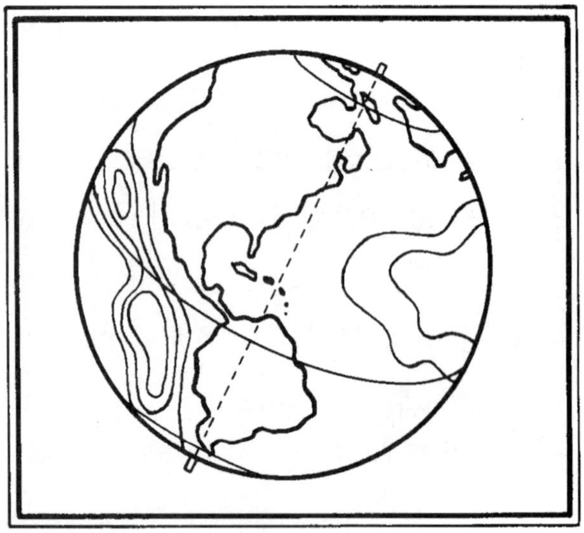

❄ Use the picture to help you answer these questions.

Why does the sun give off less heat at the Arctic Circle than at the equator?

______________________________________________

______________________________________________

What other area of the world would have similar weather conditions to the Arctic Circle?

____________________________________

 *© Teacher Created Materials, Inc.*

Name: ______________________________

# Tropical Rain Forest

For this activity you will need the Weather Alternet CD-ROM.

Click on the link to **Page 33**.

❄ Read about rain forests and answer the following.

1. Where are rain forests found? ______________________________
2. How much of the earth is covered by rain forests? ____________________
3. Find out three facts about rain forests.
   - ______________________________
   - ______________________________
   - ______________________________

❄ Look at the diagram of the rain forest. Describe the layers of the rain forest.

The first layer ______________________________

______________________________

The second layer ______________________________

______________________________

The third layer ______________________________

______________________________

Name:______________________________

# The Seasons

For this activity you will need the Weather Alternet CD-ROM.

Click on the link to **Page 34**.

This diagram shows the earth in two different positions as it rotates around the sun.

❄ Show which part of the earth is having winter and which part is having summer for each position.

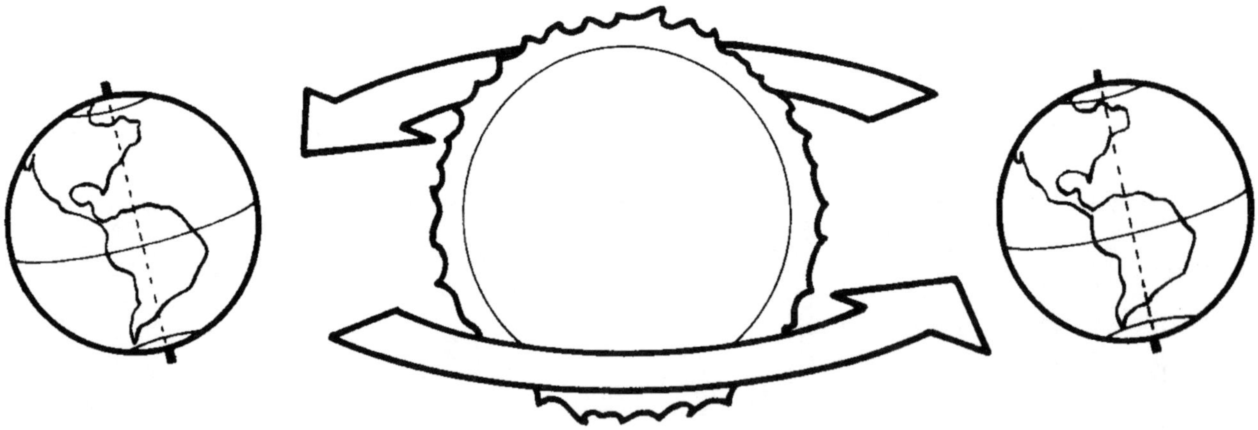

❄ Describe what happens in:

Summer ________________________________________

______________________________________________

Autumn ________________________________________

__________________________________

Winter ______________________________

__________________________________

Spring ______________________________

__________________________________

 © Teacher Created Materials, Inc.

# Water Experiments

## Content Area(s):

- earth science

## Objectives:

In this section students:

- conduct a simple experiment to show how a cloud forms.
- discuss the results of the experiment.
- explain the formation of a cloud in their own words.
- explore the insulating properties of snow through a simple experiment.
- discover how much space snow takes up compared to water.

## Materials Required:

- pencil or pen
- experiment materials as stated on activity pages
- General Activity Pages

## Time:

- approximately 20–35 minutes

## Background Information:

### Let's Make a Cloud (Page 36)

- Water is found in the air as an invisible gas called water vapor. The cloud is formed in the experiment because as the air inside the jar rises and is cooled by the ice, the water vapor it contains condenses into droplets. The droplets join together with other droplets to form the cloud. The temperature at which this happens is called the *dew point.*

### Super Snow (Page 37)

- When the air temperature falls below freezing, rain clouds produce snow instead of rain. If the temperature stays below freezing all the way to the ground, snow falls. If the temperature doesn't stay below freezing, the snow turns to rain before it lands on the earth.
- When snow falls to the ground it acts like a blanket. It stops heat from escaping and keeps the soil from freezing.
- When snow melts, it takes up less room than water because fallen snow is much lighter and less dense than water. The reason is that there is a lot of air between the flakes so when they melt, the volume of the resulting water is less than the frozen snow.

Name:______________________________

# Let's Make a Cloud!

❄ It is easy to make a cloud with the right equipment. Here's how.

You will need:

- a large glass jar
- a small metal baking tray
- some ice and some hot water

Here's what to do:

1. Pour about one inch (2.5 centimeters) of water into the jar.
2. Place some ice cubes in the baking tray and put the tray on top of the jar.
3. A cloud will form in the jar.

❄ Draw a diagram of your experiment.

❄ Discuss with your teacher and your classmates why the cloud formed.

Explain in your own words why the cloud formed.

______________________________________________________________________

______________________________________________________________________

______________________________________________________________________

______________________________________________________________________

______________________________________________________________________

______________________________________________________________________

 © Teacher Created Materials, Inc.

Name: ______________________________

# Super Snow

Snow forms high up in the air. It is made up of tiny crystals that join together to make snowflakes that fall to the ground.

❄ Do these experiments to find out about snow.

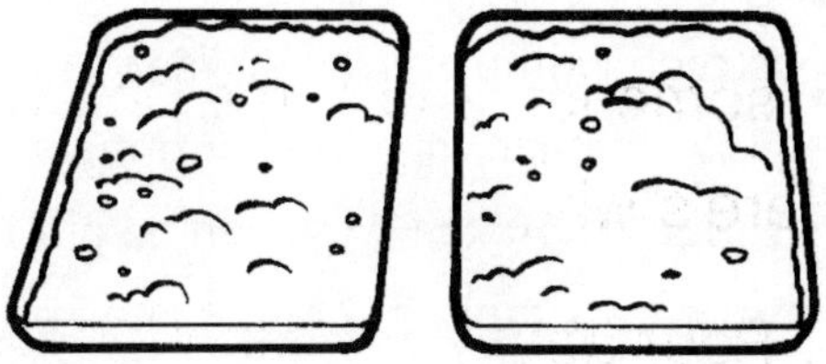

You will need:

- 2 small trays of soil
- some snow or frost from the freezer

Here's what to do:

- Make sure each tray has the same amount of soil in it.
- Cover one tray of soil with a thick layer of snow or frost.
- Put the two trays in a freezer overnight.

Which tray of soil was the most frozen? Explain why. ____________________

____________________________________________

____________________________________________

How much space does snow take up?

You will need:

- two glasses exactly the same size
- snow or frost from the freezer

Here's what to do:

- Fill one glass up with water and the other one up with snow or frost from the freezer.
- Wait for the snow to melt.

What happens? Explain. ____________________

____________________________________________

____________________________________________

# Wind

## Content Area(s):

- earth science

## Objectives:

In this section students:

- make a simple instrument to measure the speed of the wind.
- use the instrument to record the wind speed over a number of days.
- construct a simple weather vane to measure wind direction.
- use the weather vane to record the wind direction over a number of days.
- investigate how the wind can be put to work.
- construct a simple machine that is powered by the wind.

## Materials Required:

- pencil or pen
- General Activity Pages
- experiment materials as stated on activity pages

## Time:

- approximately 20–35 minutes.

## Background Information:

### Measuring the Wind 1 (Page 39)

- The pressure of the air on the earth's surface is different in different places. The differences in pressure causes the movement of air or winds. If a pressure difference exists, air moves from the higher pressure to the lower pressure area, in order to even out the pressure. If the high pressure center is close to a low pressure center, the movement from the high pressure to the low pressure is faster and the winds will move faster. The larger the pressure difference, the stronger the winds are. Instruments such as anemometers or wind gauges are used to measure the speed and strength of the wind.

### Measuring the Wind 2 (Page 40)

- The direction of the wind is determined by the direction the wind is coming from and not the direction it is blowing to. Instruments such as weather vanes and windsocks can measure the direction of the wind.

### Wind Power (Page 41)

- Using the wind for power is a good idea because it does not cause pollution. In places that are very windy, wind farms are set up where a field full of windmills turn machinery that generates electricity.

 *© Teacher Created Materials, Inc.*

Name: ______________________________

# Measuring the Wind 1

❄ You can make a simple instrument to measure the speed of the wind.

You will need:

- paper plate, wooden stick (about one foot or 30 cm long), adhesive tape, cotton reel, four paper cups (one has to be a different color from the rest)

Here's what to do:

1. Tape the cups evenly around the edge of the paper plate. The cups should all face the same way.
2. Tape the cotton reel to the bottom of the plate. Push one end of the stick into the center hole in the cotton reel and the other in the ground.

How to use the instrument:

The plates spin around as the cups catch the wind. If you count how many times the different cups go round in a minute, you can record the wind speed. Use this information to record the wind speed at different times every school day for a week.

| Day | 10 pm | 1pm | 3pm |
|---|---|---|---|
| | | | |
| | | | |
| | | | |
| | | | |
| | | | |

Which day had the highest wind speed? ________________

What time did the highest wind speed happen? __________

Which day had the lowest wind speed?________________

____________________________________________________

Name:____________________________

# Measuring the Wind 2

❄ You can find out which direction the wind is coming from by making a weather vane.

You will need:

- cardboard, knitting needle, pen top adhesive tape, compass, paper

Here's what to do:

1. Cut out a cardboard arrow and tape the top of the pen to the back of the arrow.
2. Put the knitting needle into the pen top and make sure the arrow is able to move freely.
3. Mark the compass points on a piece of square paper.
4. Place the paper on the ground and push the knitting needle through it. (Make sure the paper has been placed so the compass points on the paper are facing the right direction. You may need a compass to help.)

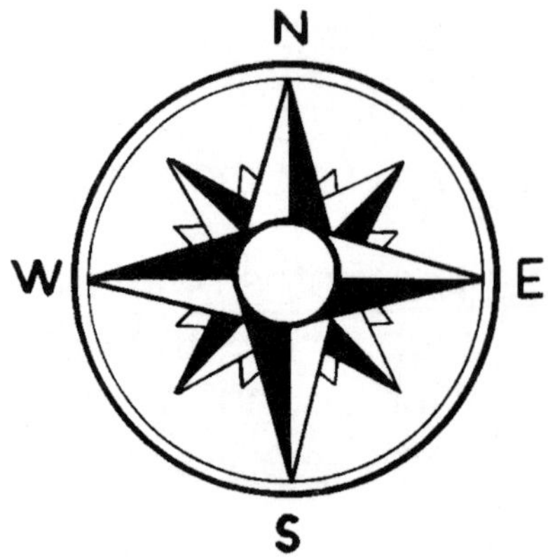


❄ Record the wind direction at different times every school day for a week.

| Day | 10 pm | 1pm | 3pm |
|---|---|---|---|
| | | | |
| | | | |
| | | | |
| | | | |
| | | | |

 *© Teacher Created Materials, Inc.*

Name: ____________________________

# Wind Power

Pollution from power stations, factories, and cars is harming the air that we breathe. Using wind power does not cause pollution.

❄ Make a windmill to show how the wind can be put to work.

You will need:

- paper, scissors, 2 straws, playdough, paper clip, adhesive tape, thread, button

Here's what to do:

1. Cut out a piece of paper about 4 inches (10 cm) square. Cut 2-inch (5 cm) slits from each corner of the square towards the center of the square.
2. Fold every second corner towards the middle and glue them down.
3. When the glue is dry, make a hole in the middle of the windmill. Push a straw through the hole and fasten it with playdough.
4. Tape a paper clip to the end of the second straw so half of the paper clip is secured to the straw and half of it forms a loop. Push the bottom end of the windmill straw through the paper clip loop so the second straw forms a handle.
5. Tape some thread on the windmill straw close to the windmill end of the straw. Tie a button to the end of the thread.
6. Hold the second straw and blow into the windmill.

What happens when you blow into the windmill? ____________________________

____________________________________________________________

____________________________________________________________

Can you think of any machines that are powered by the wind? ________________

____________________________________________________________

# Temperature

## Content Area(s):

- earth science

## Objectives:

In this section students:

- find out how a thermometer works.
- discuss water in the air.
- conduct simple experiments.
- discuss the results of various experiments.
- explore the different forms of water.

## Materials Required:

- pencil or pen
- experiment materials as stated on activity pages
- General Activity Pages

## Time:

- approximately 20–35 minutes

## Background Information:

### Hot and Cold (Page 43)

- Most thermometers have liquid (usually mercury) inside them that expands as the air gets warmer. The hotter the temperature is, the higher the liquid rises, and the temperature can be read on a scale. Other liquids such as alcohol and ether can also be used in thermometers.
- The first sealed thermometer was invented by Gabriel Fahrenheit, a German physicist who also developed the Fahrenheit scale. A Swedish astronomer, Anders Celsius, developed the Celsius or centigrade scale. Other scales have been developed throughout the years but the Fahrenheit and Celsius scales are the two most commonly used.

### Water, Water, Water (Page 44)

- The amount of water in the air is known as the humidity and is measured using a barometer. Warm air can hold more moisture than cool air.
- When a can is taken out of the refrigerator, it cools the surrounding air. The cool air cannot hold as much water. The water vapor in the air condenses and the water droplets form on the side of the can proving that there was water in the air.
- Water exists in the atmosphere in three ways: water vapor which is an invisible gas, water droplets, or ice crystals.
- The water vapor is in the form of a gas, the water droplets are liquid, and the ice crystals are solid.

 © Teacher Created Materials, Inc.

Name: ______________________________

# Hot and Cold

The temperature is how hot or cold the weather is. We measure the temperature with a thermometer. How does a thermometer work?

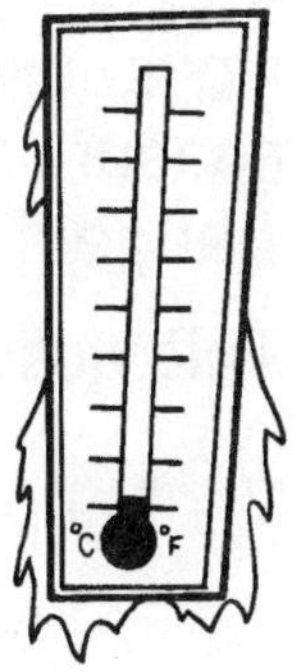

❄ Do this experiment to find out.

You will need:

- food dye, plastic bottle, straw, playdough, bowl of hot water, bowl of cold water

Here's what to do:

1. Fill the plastic bottle with water. Add some food dye.
2. Put the straw in the bottle and seal with the playdough. This is your thermometer.
3 Stand the thermometer in the bowl of hot water. Describe what happens.

_______________________________________________

_______________________________________________

4. Now stand the thermometer in cold water. What happens?

_______________________________________________

_______________________________________________

What do you think would happen if the thermometer was placed in a bowl of warm water?

_______________________________________________

_______________________________________________

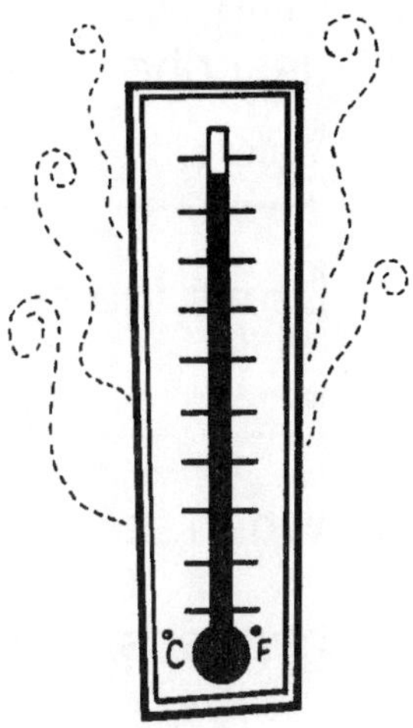

Can you explain the results? ____________________

_______________________________________________

_______________________________________________

_______________________________________________

Name:________________________

# Water, Water, Water

❄ Try these simple activities to show that there is water in the air.

1. After it has rained, measure the size of a puddle in the playground by drawing around it with chalk. Re-measure the puddle every hour to see how it has changed.

   (This activity will not work if it rains again.)

   What happens to the puddle?______________________

   ______________________

   Where does the water from the puddle go? ______________________

   ______________________

2. Leave a canned drink in a refrigerator overnight. Take the can out the next day.

   Describe what happens. ______________________

   ______________________

## Solid, Liquid, or Gas?

❄ Is water a solid, liquid, or gas? Do the following activity to find out.

Fill a container with ice. Brainstorm some words to describe water in this form (for example, hard).

______________________

Leave the container in the classroom. Describe what happens. __________

______________________

When the ice melts, put it in a kettle or a pot on a stove and boil it. What happens to the water? ______________________

______________________

 *© Teacher Created Materials, Inc.*

# Rain

## Content Area(s):

- earth science

## Objectives:

In this section students:

- conduct an experiment to make rain.
- explain the results of the experiments.
- label diagrams of the experiments.
- experiment to form a rainbow
- investigate the colors of a rainbow.
- draw conclusions from the results of the experiment.

## Materials Required:

- pencil or pen
- General Activity Pages
- experiment materials as stated on activity pages

## Time:

- approximately 20–35 minutes

## Background Information:

### Let's Make Rain! (Page 46)

- Water vapor that has been evaporated by the sun condenses on particles (specks of dust, smoke and salt from the sea). The tiny, lightweight water droplets form clouds. In warm clouds, the droplets bump together and form larger drops of water. They eventually become too heavy to stay up and the earth's gravity pulls them to the ground.
- In the experiment the heated water evaporates, becoming water vapor. When it hits the cold saucepan, the water vapor condenses and forms rain.

### Rainbows (Page 47)

- Rainbows appear when there is sun and rain at the same time and you are between the two with the sun at your back. The raindrops split the sun's rays into bands of red, orange, yellow, green, blue, indigo and violet. The colors of the rainbow are always the same and always occur in the same order.
- The shape of the rainbow is due to the way the light enters the raindrops and is refracted (bent) at certain angles. A rainbow would be seen as a complete circle if the earth's surface was not there.

Name:____________________________

# Let's Make Rain!

❄ It is easy to be a rainmaker. Do the following experiment to find out how.

You will need:

- kettle/teapot
- ice cubes
- oven mitt
- saucepan

Here's what to do:

(**Warning:** Be very careful when doing this experiment because you will be using boiling water.)

1. Boil the water in the kettle or teapot.
2. Put the ice cubes in the saucepan. Put on the oven mitt.
3. Hold the saucepan over the steam coming out of the kettle or teapot.

❄ Answer these questions about the experiment.

What happened to the water when it was heated? ____________________

____________________________________________

What happened to the water vapor when it hit the cold saucepan?

____________________________________________

____________________________________________

❄ Find out what these weather words mean:

evaporation ____________________________

________________________________

condensation___________________________

________________________________

water vapor ____________________________

________________________________

rain ________________________________

________________________________

 © Teacher Created Materials, Inc.

Name: ______________________________

# Rainbows

❄ Rainbows form on a sunny day when there has been a shower of rain. It is the sunlight hitting the raindrops that make the colors of the rainbow.

All you need to make a rainbow is a garden hose and some sunshine.

Here's what to do:

(This experiment is best done in the morning or the afternoon when the sun is low in the sky.)

- Turn on the hose and point the nozzle up high to form an arc of water. Make sure that you are standing with your back to the sun.
- Get a friend to look through the water spray. He or she should see a rainbow.

1. What are the colors of the rainbow? ______________________________

______________________________________________

2. Are they always the same? ______________________________

______________________________________________

3. Draw a rainbow showing the colors in their right order.

# Answer Key

## Snowflakes (page 7)

Snow Facts: 1. True, 2. False, 3. True, 4. False, 5. True, 6. True

## Lightning (page 9)

Sky, cumulus, larger, higher, darken, winds, particles, charged, electricity, overflow, cloud, cloud, earth, shortest, highest, flat field; ligntning takes the shortest path; true, false, false.

## Tornado Safety Tips (page 15)

Home—Get away from windows; Go to the basement; If there is no basement go to an inside closet, bathroom, hallway or lowest part of the house; Get under a mattress and protect your head.

Mall—Get off the street; Stay away from windows and doors.

Outside—Get out of a car and go inside a house or a building; Don't try to outrun a tornado in a car; If you are caught outside, lie in a ditch or crouch near a building; Cover your head with your hands.

School—Follow directions; Go to an inside hall on the lowest floor; Crouch near the wall; Bend over with your hands on the back of your head; Keep away from glass and stay out of big rooms like gyms and cafeterias.

Trailer—If you live in a trailer or mobile home, get out quickly because it can easily be shattered by the tornado.

## Weather Instruments (page 19)

1. Fahrenheit, Celsius; 2. air pressure; 3. weather balloon, rain gauge, weather satellite; 4. the temperature at which water vapor condenses; 5. weather vane; 6. anemometer; 7. by the direction it comes from

## Water (page 26)

precipitation-rain, sleet-rain that freezes on the way to the earth, atmosphere-layer of air that surrounds the earth, freezing point-32°F or 0°C, humidity-amount of water in the air, vapor-gas, gravity-pull to the earth.

### The Water Cycle (page 27)

1. sun, water, 2. water, 3. cloud, 4. clouds, 5. ground, 6. lakes, oceans and puddles, 7. water cycle.

## Desert Weather (page 31)

1. less than 10 inches; 2. near the equator; 3. The sun's rays are more concentrated near the equator. 4. Answers will vary.

## Arctic Weather (page 32)

1. the solid mass of permanently frozen earth, ice and rock; 2. up to 1,000 feet (300 meters) deep in winter; 3. It partly thaws; 4. 2 to 6 feet (0.6 to 2 meters) deep in summer.

The sun's rays are less concentrated at the Arctic Circle; Antarctica (South Pole).

## Tropical Rain Forest (page 33)

1. In the tropics close to the equator; 2. 7%; 3. Three facts: Tropical rain forests stay green all year. Because of the warm, moist climate, rain forest plants also grow very fast. A rain forest has different layers of plant growth.

## The Seasons (page 34)

Summer is the warmest season of the year. The sun can be very strong in summer and it is important to protect yourself from the sun's rays by wearing sun screen and a hat. In the top part of the world the summer months are June, July, and August. In the bottom half of the world the summer month's are December, January, and February.

In autumn the weather is not as hot as summer. The days are warm and the nights are cool. As winter approaches the air becomes colder. Some trees lose their leaves in autumn and the land is dry after the hot summer. In the top half of the world autumn is in September, October, and November. In the bottom half of the world the autumn month's are March, April, and May.

Winter is the coldest season of the year. Many areas of the world have very cold conditions. Ponds and lakes turn to ice and snow falls from the sky. In cold weather, people wear thick clothes to keep warm. Animals that live in very cold areas grow thick fur to keep out the cold. Some animals hide away in their warm burrows for the winter. Some birds fly away to warm countries and return when the weather warms up.

Spring is the season between winter and summer. The top half of the world has spring in March, April, and May. In the bottom half of the world spring weather begins about September and ends in November. The days of spring are warmer than winter days. Snows that has fallen during the winter months melt. Flowers bloom and animals leave their winter sleeping places. The days of spring have more hours of daylight and become warmer as summer approaches.

## All General Activity Pages (page 35–47)

Answers will vary.